101 Things I Learned in Engineering School

Other books in the series
101 Things I Learned in Business School
101 Things I Learned in Culinary School
101 Things I Learned in Fashion School
101 Things I Learned in Film School
101 Things I Learned in Law School
101 Things I Learned in Architecture School (MIT Press)

101 Things I Learned® in Engineering School

John Kuprenas with Matthew Frederick

GRAND CENTRAL
PUBLISHING

NEW YORK   BOSTON

Matthew Frederick is the series creator, editor, and illustrator.

Grand Central Publishing
Hachette Book Group
237 Park Avenue
New York, NY 10017
www.HachetteBookGroup.com

Printed in China

QUAL

First Edition: May 2013
10 9 8 7 6 5 4 3 2 1

Grand Central Publishing is a division of Hachette Book Group, Inc.
The Grand Central Publishing name and logo is a trademark of Hachette Book Group, Inc.

The Hachette Speakers Bureau provides a wide range of authors for speaking events. To find out more, go to www.hachettespeakersbureau.com or call (866) 376-6591.

The publisher is not responsible for websites (or their content) that are not owned by the publisher.

Library of Congress Cataloging-in-Publication Data has been applied for.
ISBN 978-1-4555-0977-5 (hardcover)
ISBN 978-1-4555-0979-9 (ebook)

From John
To my family

Author's Note

Engineers view their profession as fascinating, creative, and full of interesting challenges. Those outside engineering often regard it as repetitive, mechanical, and frustrating.

What is evident from both perspectives is that engineering is complex. It requires an intensive core of knowledge in mathematics, physics, and chemistry, the exploration of which fills most of the first two years of the college curriculum. In focusing on these elements, the curriculum tends to provide very little context. When I was a beginning engineering student, I found it frustrating that the calculations and abstract concepts I was learning in the classroom were difficult to tie to real-world application. The engineering curriculum gave me a lot of trees, and very little forest.

101 Things I Learned in Engineering School flips this around. It introduces engineering largely through its context, by emphasizing the common sense behind some of its fundamental concepts, the themes intertwined among its many specialties, and the simple abstract principles that can be derived from real-world circumstance. It presents, I believe, some clear glimpses of the forest as well as the trees within it.

It is my hope that this book will interest and enlighten college students seeking context for their developing mathematical and scientific knowledge, inspire practicing engineers to reflect on the subtle relationships in their field, and encourage the lay person to see the engineering world as engineers do—fascinating, creative, challenging, collaborative, and unfailingly rewarding.

John Kuprenas

Acknowledgments

From John

Thanks to Weston Hester, Keith Crandall, Ben Gerwick, William C. Ibbs, Povindar K. Mehta, David Blackwell, the inspiration of the books at Skylight and Powell's, and the conversations at Figaro Café.

From Matt

Thanks to Karen Andrews, Nicole Bond, Regina Brooks, Nancy Byrne, Sorche Fairbank, Venkataramana Gadhamshetty, Meredith Haggerty, Harmonie Hawley, Dylan Hoke, Dave McNeilly, Jamie Raab, Aaron Santos, Simon Schelling, Kallie Shimek, Rekha Ramani, Nick Small, Flag Tonuzi, Tom Whatley, and Rick Wolff. Special thanks to Marshall Audin, Myev Bodenhofer, and David Mallard for their ideas, help, and support.

101 Things I Learned in Engineering School

Engineering succeeds and fails because of the black box.

A black box is a conceptual container for the knowledge, processes, and working assumptions of an engineering specialty. On multidisciplinary design teams, the output of one discipline's black box serves as the input for the black boxes of one or more other disciplines. The designer of a fuel system, for example, works within a "fuel system black box" that produces an output for the engine designer; the engine designer's black box outputs to the automatic transmission designer, and so on.

Design solutions don't emerge linearly, however, and design teams work in interconnected webs of relationships. Hence, the black box model works best when employed as a momentary ideal that is adjusted and redefined throughout the design process as constraints become evident, opportunities emerge, prototypes are tested, and goals are clarified. It fails when expected to be permanent and orderly.

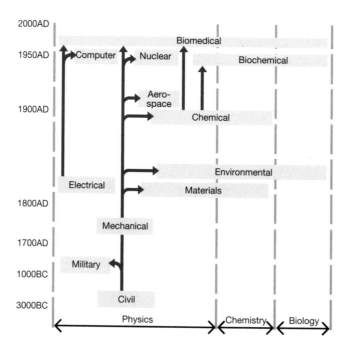

Engineering family tree

Civil engineering is the grandparent of all engineering.

In its early days during the Roman Empire, civil engineering was synonymous with military engineering. Their kinship was still strong when the first engineering school in America was founded in 1802 at the U.S. Military Academy at West Point, New York. USMA graduates planned, designed, and supervised the construction of much of the nation's early infrastructure, including roads, railways, bridges, and harbors, and mapped much of the American West.

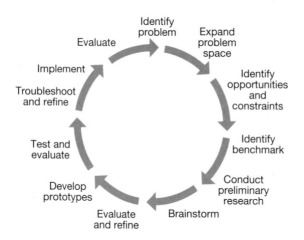

The heart of engineering isn't calculation; it's problem solving.

School may teach the numbers first, but calculation is neither the front end of engineering nor its end goal. Calculation is one means among many to an end—to a solution that provides useful, objectively measurable improvement.

Block on a ramp

Gravity component vectors

Free body diagram

Every problem is built on familiar principles.

Every problem has embedded in it a "hook"—a familiar, elemental concept of statics, physics, or mathematics. When overwhelmed by a complex problem, identify those aspects of it that can be grasped with familiar principles and tools. This may be done either intuitively or methodically, as long as the tools you ultimately use to solve the problem are scientifically sound. Working from the familiar will either point down the path to a solution, or it will suggest the new tools and understandings that need to be developed.

Straps with buckle

Out-swinging doors

How the former Hotel Louis XIV in Quebec prevented guests
from locking each other out of the shared bathrooms

Every problem is unique.

Engineering problems rely on the familiar, but invention is also called for. Some problem-solving tools are developed through rote and repetition; some emerge intuitively; some rote-learned tools become intuitive over time; and some come out of necessity and even desperation. Add the tools you develop from solving each problem to your toolbox to use on future problems. More importantly, add to your toolbox the methods by which you *discovered* the new tools.

Straight pipe
friction loss of 5.5'
per 100' of run

90° elbow
friction loss equal to
4.0' straight pipe

45° elbow
friction loss equal to
2.0' straight pipe

Tee, side outlet
friction loss equal to
8.0' straight pipe

The problem of calculating pressure loss can be simplified by
converting components to an equivalent length of straight pipe.
(Assumes 1½" dia. PVC pipe at 30 gallons per minute initial flow)

"Inside every large problem is a small problem struggling to get out."

—TONY HOARE

Result

130 lb. force

A single force vector

Result

120 lb. force

50 lb. force

Two equivalent force vectors

You are a vector.

A force is expressed graphically by a vector. A vector's length is its magnitude, and its direction is given in relation to the *x*, *y*, and *z* axes. Every person has a gravity force vector with a magnitude measured in pounds or newtons, and a direction toward the center of the earth. Any single vector can be replaced by more than one component vectors, and vice versa.

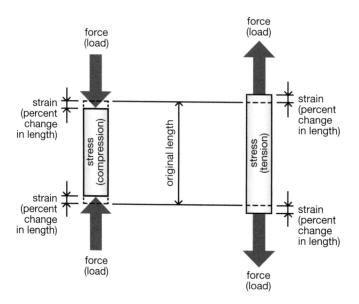

force
(load)

force
(load)

strain
(percent
change
in length)

strain
(percent
change
in length)

strain
(percent
change
in length)

strain
(percent
change
in length)

stress
(compression)

stress
(tension)

original length

force
(load)

force
(load)

An object receives a force, experiences stress, and exhibits strain.

Force, stress, and strain are used somewhat interchangeably in the lay world and may even be used with less than ideal rigor by engineers. However, they have different meanings.

A **force**, sometimes called "load," exists external to and acts upon a body, causing it to change speed, direction, or shape. Examples of forces include water pressure on a submarine hull, snow loads on a bridge, and wind loads on the sides of a skyscraper.

Stress is the "experience" of a body—its internal resistance to an external force acting on it. Stress is force divided by unit area, and is expressed in units such as pounds per square inch.

Strain is a product of stress. It is the measurable percentage of deformation or change in an object, such as a change in length.

Object remains static

Translation (sliding)

Rotation

Object moves

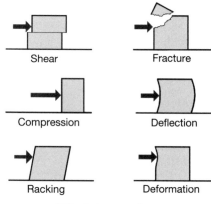

Shear	Fracture
Compression	Deflection
Racking	Deformation

Object changes shape

When a force acts on an object, three things can happen.

An object that receives a force will remain stationary, move, or change shape—or a combination. Mechanical engineering generally seeks to exploit movement, while structural engineering seeks to prevent or minimize it. Nearly all engineering disciplines aim to minimize changes in the shape of the designed object.

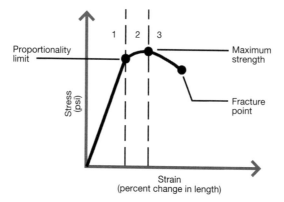

Simplified stress-strain curve

When a force acting on a fixed object increases, three things happen.

1. Proportional elongation: When an object, such as a steel bar, is subjected to a stretching (tensile) force, it initially will deform in proportion to the loads placed on it. If load x causes the bar to deform d, $2x$ will cause deformation $2d$, $3x$ will cause $3d$, and so on. If the load is removed, the bar will return to its original length.

2. Disproportional elongation: Beyond a certain point of loading (which varies among materials) an object will deform at a rate greater than the rate of increase in loading. If load $10x$ causes deformation $10d$, load $10.5x$ may cause $20d$. When the load is removed, the material will not quite return to its original length.

3. Ductility: If loading is further increased, the material will become visibly deformed and will soon fracture.

Stiff, strong, but brittle
(low-ductility) material

Less stiff, less strong, highly ductile,
ultimately tougher material

Simplified stress-strain curves

Four material characteristics

Stiffness/elasticity concerns the lengthening or shortening of a material under loading. Stiffness is resistance to change in length; elasticity is the ability to return to original size and shape. Stiffness is measured formally by the **modulus of elasticity,** which is the slope of the straight line portion of the stress-strain curve: the steeper, the stiffer.

Strength is a measure of a material's ability to accept a load. The maximum strength of a material (usually tested in tension rather than compression) is represented by the highest point on the stress-strain curve.

Ductility/brittleness is the extent a material deforms or elongates before fracturing. A highly ductile material is taffy-like, and its stress-strain curve extends far to the right. A very brittle material is chalklike; its curve ends abruptly after reaching maximum strength.

Toughness is an overall measure of a material's ability to absorb energy before fracture. It is represented by the total area under the stress-strain curve.

steel rebar
(resists tension)

back (main) span

cantilever

Steel-reinforced concrete beam

Materials fight.

Materials shrink and grow with variations in atmospheric conditions, and change in strength, shape, size, and elasticity as they age. Where different materials have similar properties, they can be intricately combined. Steel and concrete have near-identical coefficients of thermal expansion; if they did not, a steel-reinforced concrete beam would tear itself apart upon ordinary temperature change.

More often, materials are not neutral toward each other. They compete for electrons where in contact, inducing corrosion; their sizes and shapes change at different rates upon variations in temperature, humidity, and air pressure; and they respond variably to wear, tear, and maintenance. An airplane tire and the wheel on which it is seated, for example, will react in different ways to the rapid changes in temperature, air pressure, and loading experienced in ordinary use by aircraft. The integrity of their relationship will be maintained only if the system is designed to perform across the entire range of its components' physical properties.

Rockwell hardness test

Harder materials don't ensure longevity.

Monel metal is a very hard alloy of nickel, copper, and iron. It is extremely corrosion resistant and is excellent for wet applications. However, in 1915, a ship was built with a hull entirely of Monel, with the expectation of an exceptionally long life. Unfortunately, the 215-foot-long, 34-foot-wide *Sea Call* had to be scrapped after six weeks of use. The Monel hull was fully intact, but the steel frame of the ship deteriorated beyond use, from electrolytic interaction with the Monel in the saltwater environment.

Response of object

Frequency of vibrational input

Extreme response when input frequency matches object's natural frequency

Minimal response of object at other frequencies

Softer materials aren't always more protective.

Packaging engineers have found that very few packages are accidentally dropped during transport to the user, and far fewer are dropped from a height that will cause damage to most products. But while accidental impacts during transit have a low probability of occurring, every product is subject to vibrational input from the vehicles in which they are shipped. The wrong type of cushioning material can amplify vehicular vibration and cause a sensitive item to fail if the net vibration imparted to the item is at its natural frequency. An improperly designed package thus can destroy the product it is meant to protect.

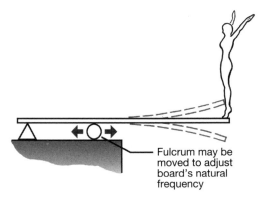

Fulcrum may be moved to adjust board's natural frequency

With each bounce, a diver stores energy in the board. By coordinating each landing with the board's natural frequency, the height of the takeoff is increased.

Soldiers shouldn't march across a bridge.

Structural members vibrate in response to normal impacts, in the manner of a plucked guitar string. The **natural frequency** or resonant frequency of an object is the time it takes to complete one cycle of movement (fully back and forth or up and down) upon disturbance.

When a force acts on a structural member repeatedly and at a rate that matches its natural frequency, the member's response is enhanced on every cycle. The effects range from loud humming (such as when vibrations from mechanical equipment coincide with a beam's natural frequency) to uncomfortable oscillation to occasional collapse. Many relatively small earthquakes have induced significant damage when their wave frequency has matched that of affected buildings. In 2000, thousands of pedestrians celebrating the opening of the London Millennium Footbridge inadvertently induced oscillation when their walking rhythms matched the structure's natural frequency. As they swayed in response to the unanticipated movement, they inadvertently increased the harmonic resonance. The bridge was closed following the event and the structural system was repaired.

Aeroelastic flutter

Why Galloping Gertie collapsed

The Tacoma Narrows suspension bridge in Washington State was the third longest of its kind when completed in 1940. Despite unusual levels of motion observed during construction, "Galloping Gertie" was opened to the public in July. In November it began to heave violently as a lone driver, Leonard Coatsworth, drove across it with Tubby, a cocker spaniel. Unable to continue, and unable to remove the dog from the car, Coatsworth fled on foot. Several unsuccessful attempts were made to rescue Tubby, but eventually the dog, car, and bridge fell into Puget Sound.

The cause of the collapse is still debated, but the Washington State Highway Department determined that it was not the result of rhythmic wind gusts "exciting" the structure's natural resonance, as is often thought. Rather, the disaster began with **aeroelastic flutter** (a vibrational response to air movement, common in airplane wings), leading to **torsional flutter** (repetitive twisting). The 2,800-foot-long, 39-foot-wide main span was particularly wind vulnerable, as its main girders were made of solid plates that were only 8' deep. By comparison, an earlier proposal for the bridge suggested a 25' deep, open web stiffening truss.

Ten years after the collapse, a replacement bridge was built. "Sturdy Gertie" incorporated the original approach ramps and main piers, but used a 33' deep, open web stiffening truss.

Anode
(more active)

Magnesium
Zinc
Aluminum
Steel or iron
Lead
Nickel
Brass
Copper
Bronze
Stainless steel 304
Monel metal
Silver
Gold
Platinum

Cathode
(less active)

Partial galvanic series

A battery works because of corrosion.

On the surface of all metals are loosely bound electrons. When two different metals are placed in contact, the atoms of each compete to attract the electrons. The more "noble" metal (**cathode**) attracts electrons from the more "active" metal (**anode**). This movement of electrons causes the anode to corrode, and produces an electric current. The common household battery generates current with a carbon-zinc cell, in which the zinc corrodes preferentially to the carbon.

starting
water
level

weight of water
displaced by
object
=
upward buoyancy
force on object

Archimedes Principle

Buildings want to float.

Buoyant uplift force on an object equals the weight of the water that the object displaces. If a building's lower floors extend below the groundwater level, buoyancy will seek to lift it—even if the displaced water is dispersed throughout the soil. Floatation is unlikely in a completed building, but a deep foundation, basement, or underground parking garage will seek to float to the surface of the earth without the weight of the building pressing down on it. For this reason, an underground storage tank must be attached to a concrete mass at least equal to the weight of the groundwater the tank will potentially displace.

air flow

low-pressure
zone

air flow

movement of wing

air flow

low-pressure
zone

movement of car

Automobiles want to fly.

A wing does not have to be shaped as an airfoil to allow an airplane to fly. A completely flat wing placed at the proper **angle of attack** will create lift as the plane moves forward by producing a low pressure zone immediately above the wing. However, an airfoil shape has far less drag and functions much more efficiently.

A low-pressure zone is similarly created through the forward motion of automobiles, particularly those with a traditional sedan shape. As a sedan moves forward, a low-pressure zone is produced above the trunk. Above about 70 mph, rear-end lift can noticeably affect a driver's control. Approaching 200 mph, a car may become airborne.

air accelerates in
narrower part of tunnel,
reducing pressure

After a drawing from U.S. patent 4,386,801 by Colin Chapman, et al.

The ground effect

The rear wing on a race car counteracts lift by introducing downforce at the rear of the vehicle. However, it increases drag, reducing aerodynamic efficiency.

British inventor Colin Chapman sought an alternative to the wing. He ultimately created a front-to-rear air channel on the vehicle's underside. The top of the channel is shaped similar to an inverted airfoil. Combined with very low ground clearance and side skirts, air moving under the vehicle is directed through a narrow zone, which causes the air to accelerate. Because faster moving air naturally has lower pressure, the vehicle is thereby "sucked" to the road. Chapman's "ground effect" proved so effective when introduced by his Lotus team into Formula One racing that it was quickly banned.

There was a trade-off, however. If a ground-effect car traveling at high speed was bumped, the air channel could be disrupted, resulting in catastrophic loss of control. But the genius of Chapman's invention is undeniable, for he inverted the perplexing problem of creating more "push-down" on top of a vehicle: he worked from the opposite side and created more "pull-down."

Conventional intersection
32
1300–1500
up to 55mph
up to 90°

Vehicle conflict points (•)
Lane capacity per hour
Operating speed
Angle of collision

Roundabout intersection
8
1800
15–25mph
low/glancing

A roundabout is the safest, most efficient intersection.

21

Where roundabouts have replaced conventional intersections, traffic delays have been reduced up to 89%, accidents 37 to 80%, injuries 30 to 75%, and fatalities 50 to 70%. Up to an 8-fold return on investment due to accident reductions has been indicated.

A civil engineering team at the University of Texas found that crossings with flashing lights are the most dangerous intersection type, with an accident rate approximately 3 times greater than stoplight intersections and 5 to 6 times greater than traffic circles.

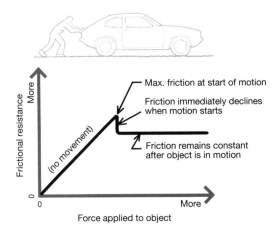

Max. friction at start of motion

Friction immediately declines when motion starts

Friction remains constant after object is in motion

Frictional resistance

More

0

(no movement)

0

More

Force applied to object

22

Friction is the enemy of a rolling object, but it is what allows it to roll.

An object is slowed by friction between it and the surface on which it slides or rolls, as microscopic bumps and ridges on each catch on the other. The greater the friction, the more a wheel's efficiency is reduced and the more heat is produced. The lesser the friction, the more freely and efficiently the wheel rolls. This suggests that a state of zero friction would be ideal for a wheel, because it would roll with perfect efficiency. But the wheel would not roll at all because of an absence of traction; rather, it would slide.

$\pi = \textbf{3.14}$
Accurate and imprecise

$\pi = \textbf{3.1415926535}$
Accurate and precise

$\pi = \textbf{3.4566289441}$
Inaccurate and precise

23

Accuracy and precision are different things.

Accuracy is the absence of error; **precision** is the level of detail. Effective problem solving requires always being accurate, but being only as precise as is helpful at a given stage of problem solving. Early in the problem solving process, accurate but imprecise methods, rather than very exact methods, will allow consideration of all reasonable approaches and minimize the tracking of needlessly detailed data.

24

There's always a trade-off.

Lightness versus strength, response time versus noise, quality versus cost, responsive handling versus soft ride, speed of measurement versus accuracy of measurement, design time versus design quality…it is impossible to maximize the response to every design consideration. Good design is not maximization of every response nor even compromise among them; it's optimization among alternatives.

Reality

Analog recording
omits some information
from reality

Digital copy
omits some information
from analog recording

Digital recording
omits some information
from reality

Digital copy
100% accurate to
digital recording

25

Quantification is approximation.

Engineering follows the laws of science, but nature does not. As a system of understanding created by humans, science is contained within reality. Nature follows itself; science is our remarkable but imperfect attempt to explain it. Quantification is exact not unto reality, but unto itself.

200km measuring unit
coastline = 2,400km

100km measuring unit
coastline = 2,600km

50km measuring unit
coastline = 3,100km

UK Ordnance Survey
coastline = 17,820km

As the increment of measurement is decreased in size, the perimeter
measurement of an irregular object increases and approaches infinity.

26

Random hypothesis #1

You don't fully understand something until you quantify it. But you understand nothing at all if all you do is quantify.

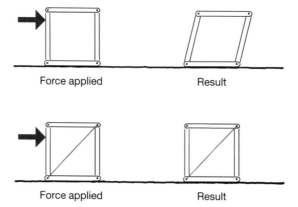

Force applied

Result

Force applied

Result

A triangle is inherently stable.

A triangle differs from other linear shapes in that its sides and angles are interdependent: a change cannot be made to an angle without altering the length of at least one side, and vice versa. By comparison, a square can be deformed into a parallelogram without changing a side.

Arch

Cantilever

truss

beam

Simple span

Tower/tensile structure

Four ways to span

28

A truss's complexity is the product of simplicity.

A truss is a complex form of beam that takes advantage of the inherent stability of the triangle. By starting with a triangle and adding two legs at a time, a series of interdependent triangles form a stable structure capable of spanning long distances, using a fraction of the material used by an ordinary beam.

Dead load: weight of the building

Live load: weight of contents

1. If total load supported by column is 240,000 lbs. and
2. Soil capacity is 3,000 lbs./sq.ft.

3. Area of bottom of footing must be at least 240,000 ÷ 3,000 = 80 sq.ft.

The contents of a building might weigh more than the building.

29

Dead load is the weight of a building itself and is constant over the life of the building. It includes the structure (beams, columns, joists, etc.), primary building systems (exterior wall, windows, roofing, interior finishes, etc.), permanent architectural elements (stairways, partitions, and flooring materials, etc.), and mechanical systems (heating, cooling, plumbing, electrical, etc.).

Live loads change over the life of a building. They come from people, furniture, vehicles, wind, earthquakes, snow, impacts of foreign objects, and similarly variable sources.

Total load is transmitted to the building foundation, then to the earth. Total load per unit area of the footing cannot exceed the bearing capacity of the soil.

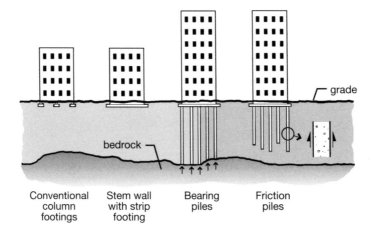

Common foundation/footing types

Structural systems are built from the bottom up, but designed from the top down.

30

The lower structural elements of a building support the elements above them. Before any structural element is designed, the elements above it must be analyzed and designed. Don't perform structural design in one top-to-bottom pass, however. Make numerous passes of increasing rigor and precision, from schematic through final design, before deciding how to best transfer loads downward to the earth.

gravity
load

wind
load

Beam cantilevered
from a wall

Skyscraper "cantilevered"
from the earth

31

A skyscraper is a vertically cantilevered beam.

The primary structural design challenge of a skyscraper is not resistance to vertical (gravity) loads, but resistance to lateral loads from wind and earthquakes. For this reason, tall structures function and are designed conceptually as large beams cantilevered from the ground.

Diagonal bracing
distributes lateral forces
among structural
members

Shear wall
extra stiff construction
resists lateral forces in the
direction of its surface

Floor diaphragm
extra stiff construction resists
horizontal forces across the
plane of the floor

Three ways of increasing lateral strength

Earthquake design: let it move a lot or not at all.

32

Earthquakes are typically characterized by lateral (side to side) movement. A structure can resist this force by being either very flexible or very rigid. In a **flexible structure**, beam-column connections rotate with relative freedom when stressed, with perhaps some dampening or diagonal shock absorbers. A **rigid structure** relies on very strong connections among structural members, with isolators (essentially, large rubber donuts) at the building base. In both systems, seismic energy is dampened so that building inhabitants experience a fraction of the earthquake's forces.

slab

joist

beam

column

foundation
(pier)

foundation
(footing)

earth

Make sure it doesn't work the wrong way.

33

The downward transfer of structural forces through a building is the **load path**. Loads sometimes follow a path different from the one intended. For example, when a non-structural partition is built under a structural beam, the beam may sag under normal loading, transferring loads to the partition, which may transfer the loads to the floor below, causing the floor to sag and even fail.

Figuring out how to make a system work is as important as figuring out how to make it not work in undesirable ways.

Rod hung
from roof

DESIGN
INTENT
4th floor beam supports
4th floor walkway

Nut secures
beam to rod

Rod supports 2nd
floor walkway below

Rod hung
from roof

UNRECOGNIZED
CHANGE
4th floor beam supports
4th floor and 2nd floor

Nut secures
beam to rod

Rod supports 2nd
floor walkway below

As designed

As built

Kansas City Hyatt walkway collapse

On July 17, 1981, two interior atrium walkways collapsed during a dance party at the Hyatt Hotel in Kansas City, Missouri. One hundred fourteen people were killed and more than 200 were injured.

The walkways crossed the atrium at the 2nd and 4th floors, and were suspended from the roof by steel rods. The engineer intended the rods to be continuous, with the upper walkway held in place by nuts and the rods continuing to the walkway below.

During construction, the fabricator recognized the difficulty of installing four-story-long threaded rods, and of rotating the nuts two stories into place. A proposal for two sets of shorter rods was put forth. One set would hang the 4th floor walkway from the roof, and a second set would hang the 2nd floor walkway from the 4th floor walkway. The engineer approved the design change without performing a structural analysis.

Analysis after the accident revealed that the revision had doubled the load on the 4th floor steel beams. Further, each beam specified by the engineer was not a single member, but two parallel members welded together. Under full loading, the welded seams of the 4th floor beams quickly failed, causing the upper walkway to "pancake" onto the lower walkway.

	Douglas fir	Concrete	A36 Steel
Maximum strength determined in laboratory testing	7,430 psi (compression)	4,000 psi (compression)	50,000 psi (tension)
Design strength used in calculations	1,350 psi	3,000 psi	36,000 psi
Approximate safety margin	5.5	1.3	1.4

Engineers wear both a belt and suspenders.

Structural systems have safety factors built into them in several ways. All materials are laboratory tested to determine their structural properties, such as the amount they stretch and compress under loading and the maximum load they will accept before failing entirely. A lower **design strength** (the presumed strength used by an engineer in performing structural calculations) is then established to allow for variability in materials used in construction.

Manufactured materials such as concrete and steel are of comparatively uniform quality, and variability from piece to piece is small. A wood beam used in a building, however, might have an unusual number of knots, have come from a diseased tree, or have reacted atypically to drying. Consequently, the design strength for wood is much lower than that determined in laboratory testing.

Engineers commonly build in additional safety margins by overestimating live and dead loads, rounding calculations to the conservative side, and selecting a structural member larger than that calculated.

The best beam shape is an I—or better yet, an I.

When a beam bends under loading, the top shortens (is compressed) and the bottom is stretched (tensioned). Compressive and tensile stresses are respectively greatest at the very top and bottom of the beam, and they decrease toward the middle, where stresses are zero at the **neutral axis**.

A rectangular beam used in a vertical orientation (B) is more efficient than the same beam used horizontally (A) because a greater percentage of its material is located where stresses are greatest—away from the neutral axis. This puts most of the material where most of the work needs to be done. For the same reason, an **I**-shaped beam (C) is even more efficient.

x
span

Simple span with
beam of depth d.

d

x
back span

1/3 x
canti-
lever

Most beams can
be cantilevered
a short distance
without increasing
depth.

d

3/4 x
back span

1/4 x
canti-
lever

A beam of a given
length can be made
shallower by moving
one or both columns
to create a backspan
and cantilever.

<d

Get even more out of a beam.

A useful measure of a beam's efficiency is the ratio between the distance it spans and the amount of material it uses. It is almost always better to use the smallest amount of material possible.

Use cantilevers. Extending a beam beyond its support helps spring the main span (or back span) upward, and often allows a smaller beam to do the same work. Most beams can be cantilevered about 1/3 the length of the main span without an increase in depth.

37

Convert point loads to multiple loads. A beam more easily resists a load if the load is distributed in several locations rather than at one point. Moving a point load from the center of a span toward one end also helps; loads from high in a building sometimes can be redistributed this way to reduce the size of beams below.

Use a truss. Although it must be deeper (taller) than a solid beam to do the same work, a truss uses a fraction of the material.

Cut holes in it. Under many loading conditions, material can be removed from the web of an I-beam, which not only lightens it but allows for the passage of ducts, pipes, and wires.

Castellated beam

"Inventing is the mixing of brains and materials. The more brains you use, the less materials you need."

—CHARLES KETTERING

38

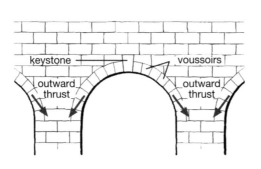

A masonry arch gets stronger as it does more work.

Gravity is usually the enemy of structural endeavor, as it seeks to pull structures toward the earth. But a masonry arch works *because* of gravity. Gravity pulls each masonry unit into contact with the unit below it, which transmits the force into the unit below it, and so on. The greater the loads on an arch—at least until the compressive strength of the material is exceeded—the greater the cohesion of the masonry units. For this reason, a masonry arch will tend to destabilize when there is a relatively small load on it, and will often look unstable, if not *be* unstable, when there is relatively little masonry above it.

At its base, an arch, in addition to a vertical gravitational force, generates an outward thrust. This must be resisted by a large mass, such as a concrete or earthen embankment for a bridge arch or a "flying" buttress for a large church arch. For an ordinary window or door opening in a masonry wall, the adjacent wall usually suffices to resist the outward force, unless the opening is very near an edge or corner. When arches are placed in series, the outward thrust from each arch resists the thrust from the adjacent one.

39

Water barrier

Air barrier

Insulation barrier

Vapor barrier

Drainage cavity

Interior finish

Wall sheathing

Studs

Residential wall

The four eras of the wall

Great Mass Era: From early civilization into the late 19th or early 20th century, very thick stone, brick, logs, adobe, and concrete walls provided the primary defense against heat, cold, wind, earthquakes, and intruders, as well as primary structural support in buildings.

Curtain Wall Era: Upon the emergence of iron, steel, and concrete structural systems in the late 19th century, and increasing specialization in building design and construction, the exterior wall shed its load-bearing role and existed primarily to enclose interior space.

40

Insulation Era: The invention of fiberglass insulation in 1938 allowed a four-inch wall to provide protection against heat and cold similar to a 2-foot thick masonry or adobe wall. But as we added more insulation to buildings and sealed them more tightly against the elements, they often developed problems with condensation and air quality.

Specialized Layers Era: The state-of-the-art building envelope today has four differentiated layers that perform as an integrated system: **water barrier** (a cladding and/or membrane that protects against precipitation and ground water), **air barrier** (a membrane that minimizes seepage of exterior air into the building), **insulation barrier** (thermally separates interior and exterior), and **vapor barrier** (a membrane that prevents movement of moist indoor air into wall or ceiling cavity). Each layer is ideally continuous around a building.

The first building code

"If a builder has built a house for a man, and has not made it sound, and the house falls and causes the death of its owner, that builder shall be put to death. If it is the owner's son that is killed, the builder's son shall be put to death. If it is the slave of the owner that is killed, the builder shall give slave for slave to the owner of the house. If it ruins goods, the builder shall make compensation for all that has been ruined, and shall re-erect the house from his own means. If a builder builds a house, even though he has not yet completed it; if then the walls seem toppling, the builder must make the walls solid from his own means."

—THE CODE OF HAMMURABI
by King Hammurabi of Babylonia, 1792–1750 BCE

41

Provide .08 thick EPDM roofing membrane, mechanically fastened. Acceptable manufacturers include Carlisle SynTek.

Provide roofing membrane adequate to keep out all elements under all conditions for at least 20 years.

Prescriptive specifications
give detailed characteristics of products or systems to be used, including materials, dimensions, and methods of installation.

Performance specifications
identify the desired performance, such as strength, capacity, and stability, without indicating how the contractor is to achieve it.

Drawings explain only some things.

Drawings, no matter how detailed, tell only part of the story of what is to be manufactured or built. A separate, comprehensive **specifications** document, also prepared by the engineer, provides details on the **what** (e.g., strength of concrete and steel, acceptable fastener types, wiring gauges), **who** (qualifications of sub-contractors, acceptable component manufacturers), **when** (schedules, sequence of tasks, procedures for review of the work), **where** (parts of a construction site that may be used for certain activities), and **how** (handling of materials, methods of applying finishes, etc.).

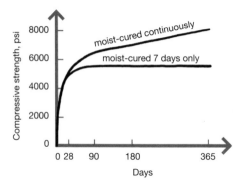

42

Concrete doesn't dry; it *cures*.

Concrete gains strength through a chemical reaction between cement and water. After pouring, concrete is often kept wet for an extended period to prolong the chemical reaction (thereby strengthening the product), and to keep outer portions of the concrete from drying long before interior portions (thereby minimizing cracking). Design calculations for concrete construction typically are based on the strength expected after 28 days of curing. However, the maximum strength of a very large pour might not be achieved for decades.

43

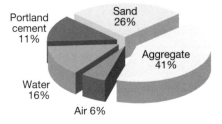

Typical concrete mix design

Concrete and cement are different things.

44

Cement is a hardening ingredient in concrete, and is usually derived primarily from limestone. Cement is mixed with sand, aggregate (rocks or pebbles), air, and water to make concrete. Chemicals also may be introduced into the mixture to speed or slow the hardening of the concrete, make it heavier or lighter, and enhance resistance to environmental factors.

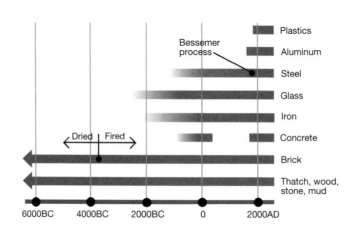

Concrete and steel are ancient, not modern, materials.

45

It was known many centuries ago that a small amount of carbon made iron into a stronger metal, but mass production of steel was not possible until the invention of the Bessemer process in the 1850s. Concrete was used in the Roman Empire, but after its dissolution the technology was lost until the 18th to 19th centuries.

Separating
Material is sized and shaped by removing excess

Casting/molding
A molten or liquid material is placed in a mold to solidify

Forming
Material is shaped via a die (a shaped metal block)

Conditioning
Material properties are changed through heat, pressure, or chemicals

Assembling
Individual pieces are combined, e.g., on an assembly line or in a garment factory

Finishing
Surfaces are protected or beautified by tempering, coating, decorating, etc.

Secondary processing

3 stages of manufacturing

Material extraction: Raw materials, such as trees, crops, oil, and minerals are identified and harvested.

Primary processing: Extracted materials are put into standard formats for use by industry. Limestone, sandstone, and shale are baked and crushed into powder to make cement; alumina is extracted from bauxite ore, processed, and cast into aluminum ingots; cotton is cleaned, deseeded, and compressed into bales; grain is ground into flour.

Secondary processing: Primary industrial materials are made into products for use by consumers.

46

Actual Condition

	Part is defective	Part is not defective
Part judged defective	correct assessment	**Type 1 error** false positive
Part judged not defective	**Type 2 error** false negative	correct assessment

Judgment by inspector

More inspections and fewer inspections both produce more errors.

Inspection occasionally rejects a good item or fails to identify a defective item. A **false positive error** has little consequence other than the cost of replacing the item. But a **false negative error** can have great consequence, as the item may fail after being placed in service.

More inspections are not necessarily the answer, however. Statistically, the addition of an infinite number of layers of inspection will cause nearly every item to be found defective for some reason. The optimal level of inspection balances the economics of replacing false positives with the human and moral consequences of failing to detect real errors.

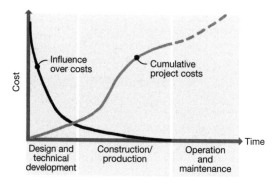

47

Early decisions have the greatest impact.

Decisions made just days or weeks into a project—assumptions of end-user needs, commitments to a schedule, the size and shape of a building footprint, and so on—have the most significant impact on design, feasibility, and cost. As decisions are made later and later in the design process, their influence decreases. Minor cost savings sometimes can be realized through **value engineering** in the latter stages of design, but the biggest cost factors are embedded at the outset in a project's DNA.

48

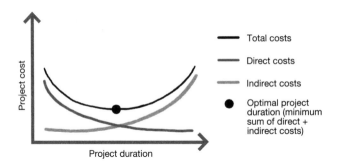

Total costs

Direct costs

Indirect costs

Optimal project
duration (minimum
sum of direct +
indirect costs)

Project cost

Project duration

Working faster doesn't save money.

When a production schedule is accelerated, savings are often expected through reduced **indirect costs**—overhead, equipment rental, insurance, supervision, utilities, and so on. Meanwhile, **direct costs** (generally, labor, materials, and equipment) are expected to stay constant, because the same amount of work must be done regardless of schedule.

In practice, however, faster work produces more confusion, errors, substandard quality, and overtime pay, ultimately increasing cost. An extremely long work period also increases total costs, and particularly indirect costs. **Optimal project duration** minimizes the combination of indirect and direct costs.

Occasionally, the greater costs of an accelerated production schedule are acceptable. In a highly profitable real estate market, a developer eager to make a building available for leasing might employ **fast-tracking**, in which construction begins before the building is fully designed. This drives up costs for the above reasons, and also because many parts of the building, such as the foundation and structural system, must be overbuilt to allow for the worst-case outcome of design decisions not yet made.

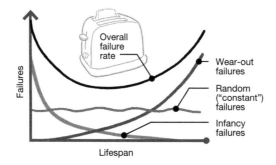

49

Common "bathtub curve" of reliability for many products

Perfect reliability isn't always desirable.

Reliability is the measure of how long a product or system functions properly. A **target reliability of one** is a goal of perfect reliability; zero indicates all failures. The target reliability of a bridge, spacecraft, pacemaker, or similarly critical system is one, because failure may result in loss of life. Products that are relatively inexpensive, such as a toy or DVD player, are designed to have less than perfect reliability because failure is not critical and achieving perfection would increase cost. Surprisingly, some aircraft parts have a target reliability of less than one because of the need to minimize weight. This is mitigated by routine replacements and frequent inspections to identify cracks and other potential failures.

Failures tend to occur for different reasons over the life of a product or system, with ordinary wear-out failures overtaking start-up or infancy failures over the long run.

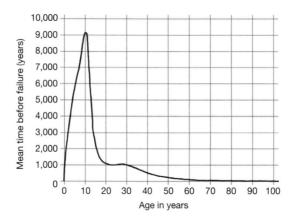

50

Human time to failure is 1,000 years.

Mean Time Before Failure is the inverse of the expected failure rate of a device or system. A 25-year-old person has a MTBF of about 1,000 years, because the annual rate of death (failing) for a person that age is 1 in 1,000, or 1/1000. As we age and near the end of our **service life**, our MTBF decreases. There is no direct correlation between service life and failure rate. A rocket is designed to have a MTBF of several million hours, because failure would be critical. However, it is intended to last only a few minutes, such as during the launch of a space ship.

The quality-cost curve

Few customers will pay for a perfectly engineered product.

Customers notice and are willing to pay for improvements to low quality products more than high quality products. A 10% improvement to a low quality product will lend more than a 10% increase in the **value of quality**—the user's perception of its quality. But as subsequent improvements are made, they add value at a decreasing rate. If a 10% quality improvement costs $10, a 20% improvement will cost more than $20. Eventually, the **cost of improving quality** increases at a faster rate than the improvement will be perceived.

The **optimal quality-cost state** theoretically occurs when the slopes of the value and cost curves are equal. At this point, the rate of improving a product equals the rate at which costs to the producer will increase. Beyond this point, the producer's cost for providing one more unit of quality will exceed the value the customer will perceive.

AMC Gremlin	294.5
Chevrolet Vega	299.0
Ford Pinto	**310.0**
Toyota Corolla	313.0
Datsun 510	317.0
VW Beetle	374.0
Datsun 1200/210	405.0

Average of group: 330.4

0   100   200   300   400

Average annual occupant fatalities per million vehicles, 1975–76

The Ford Pinto wasn't unsafe.

Following the influx of small foreign cars into the U.S. in the 1960s, Ford developed the Pinto on an accelerated schedule. Shortly after its release, the vehicle was assailed for catching fire in rear-end crashes. Over 500 deaths were said to have resulted from design flaws such as bolts protruding from the rear differential near the gas tank.

In a wrongful death lawsuit, an internal Ford document surfaced that stated that unimplemented improvements to the gas tanks would have cost only $11 per vehicle. Using a human life value of $200,000, Ford determined it would cost far less to pay for injuries and deaths than to improve 12.5 million vehicles. The legal standard of the day was expected to excuse Ford from liability, because the courts did not consider a defendant negligent if the cost of an improvement exceeded its benefit. But the jury found Ford liable and ordered it to pay $3M compensation and $125M in punitive damages (later reduced to $3.5M).

A later study revealed that the unimplemented $11 improvement was never meant to address gas tank failures in rear-end collisions. Nor did Ford place a $200,000 value on human life; this value was created by the National Highway Traffic Safety Administration. And statistics suggest the Pinto's overall safety record was about average for its day, with its registration rate matching its rate of involvement in vehicle fatalities. Nevertheless, perceptions of the Pinto's danger and of Ford performing a questionable moral calculus have lived on.

53

IKEA furniture frequently employs one set of hardware to align parts, and a separate set of hardware to fasten the parts together. Each set serves one purpose, thereby minimizing opportunities for error by the inexperienced home assembler.

Hidden clips for installing porch boards: Clips serve two purposes, but only one purpose at a time. On one side of a board, they are fastened to the structure below. On the other side, they are tucked under the previously installed board. If two clip types were provided, one could easily install the wrong clip.

Be careful when asking a part to do more than one thing.

It may seem desirable to minimize effort, material, and time by having one feature or part serve multiple purposes. However, this depends on the level of skill and care that can be expected during application. The greater the sophistication of the end user and the more controlled the user's environment, the more one may rely on multi-functionality. But where an error would be catastrophic, it is usually better to have each part serve only one purpose.

54

Design a part to fail.

Electrical systems are protected by fuses or circuit breakers that blow or trip before power surges can ruin expensive components or damage hard-to-access wires.

In **steel buildings**, connections between structural members may be designed to deform during earthquakes, to prevent catastrophic failure of the larger system. Repair of the connections can be done at a fraction of the cost of replacing the entire building.

Biomedical devices are often connected loosely to protect a patient. A pulse oximeter, used to detect blood oxygenation, is connected to a patient's finger with deliberate weakness, preventing injury should someone trip over the cable.

The clips that hold **lobster traps** together are designed to corrode after one fishing season. When traps are lost or abandoned, the clips will fail before the wire-gridded sides, leaving flat pieces that are much less hazardous to boats than a pile of submerged boxes.

up to 1mA	Tingle sensation	(1000 mA = 1 Amp)
1-2 mA	Uncomfortable sensation	
5 mA	Maximum harmless current	
15 mA	Maximum "let-go" current	
10-20 mA	"Can't let go" current	

Electric shock values

Keep one hand in your pocket.

If one of your hands is touching any object while the other hand touches a piece of electrical equipment that releases an errant charge, the charge may seek a path to the ground, from one hand to the other, through your heart. Keeping one hand in your pocket won't keep you from getting shocked, but it will encourage a charge to follow a less dangerous path—through your hand, up your arm, and down the nearest leg to the ground.

56

Keep one leg still.

To level a surveying tripod, set it in the desired location at approximate level. Repeatedly adjust two of the legs and ignore the third leg until the bull's-eye indicates dead level.

57

How to read a topographic plan

A topographic plan depicts a landscape through a series of contour lines. Each line indicates a constant **elevation**—a measured height above sea level or other reference point. Several keys help in reading a topographic plan:

- The direction of slope is always perpendicular to the contours. A meandering drop of rainwater moves perpendicular to the contour lines, from higher to lower elevation.
- Where contour lines are more closely spaced, the terrain is steeper; where farther apart, shallower.
- If a nearby lake were to flood to a given elevation, the outline of the lake would match a contour line.
- If discerning a ridge (or crown) from a gully (ditch) is difficult, place yourself on the plan at the edge of the ridge or gully, and imagine yourself walking straight across it. At each step, verify the elevation on the plan to determine if you are going up or down.

Existing

Proposed

Balance cut and fill.

Design sitework to equalize the amount of earth to be removed (**cut**) and the amount to be added (**fill**). This simplifies earthmoving and grading, and minimizes the expense of moving soil to and from the construction site.

59

The Panama Canal, schematic section looking northeast

Work with the natural order.

The locks of the Panama Canal are operated without any pumps. Gravity moves millions of gallons of water from lakes to the lock chambers, where ships are raised and lowered 85 feet in passing between the Atlantic and Pacific Oceans. As long as precipitation refills the lakes, the locks continue to function.

60

Turbulent flow
Particle paths are irregular. Tends
to occur in larger pathways and at
high flow rates

Laminar flow
Particles move in straight lines.
Tends to occur in small pathways
and at low flow velocities

Two types of flow in fluids

Air is a fluid.

A fluid is any amorphous substance that yields easily to external pressure and assumes the shape of its container. This includes all gases and liquids.

61

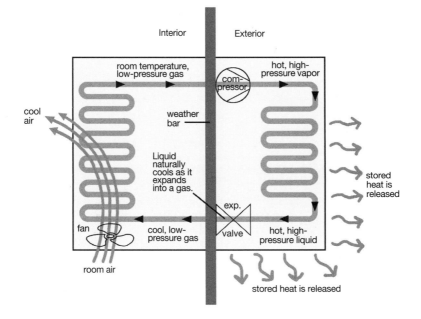

Interior | Exterior

room temperature, low-pressure gas

com-pressor

hot, high-pressure vapor

cool air

weather bar

Liquid naturally cools as it expands into a gas.

exp. valve

stored heat is released

fan

cool, low-pressure gas

hot, high-pressure liquid

room air

stored heat is released

An air conditioner

Heat cannot be destroyed, and cold cannot be created.

An air conditioner doesn't create cold, but moves heat from a building interior to the exterior. It does this by exploiting a natural principle: substances absorb heat when moving from a liquid phase to a gas phase, and release heat when moving from gas to liquid. Central air conditioning systems, window air conditioners, and food refrigerators work the same way but at different scales. A heat pump is conceptually an air conditioner working in the opposite direction, removing heat from outdoor air and moving it into the building interior.

Conduction
heat transfer through
direct material contact

Convection
heat transfer through movement of a fluid (air or liquid)

Radiation
energy transfer
through space

A radiator doesn't just radiate.

Heat is the movement of molecules within a material. The greater the rate of movement, the greater the heat. Heat is transferred by:

Conduction: When two objects of different temperatures are in contact, or when two areas of one object are at different temperatures, the more active molecules in warmer areas "nudge" the molecules in the colder areas until all molecules are moving at the same speed.

Convection: In gases and liquids, the molecules of the warmer material naturally spread out and move throughout the colder material, losing their heat energy to it. In this way a radiator convects, not just radiates, heat to a room.

Radiation is the movement of electromagnetic waves through space, such as light waves from the sun. The waves provide energy to the molecules that they contact, causing them to become more active, and converting electromagnetic energy into heat energy. All matter emits thermal radiation; most can't be felt.

A double envelope house with earth-derived heating and cooling

The most reliable source of heating and cooling is the earth.

Several feet below the surface of the earth, the temperature is more moderate than the air temperature—cooler in the summer and warmer in the winter. In the same way an air conditioner transfers heat from a building interior to the exterior air, or a heat pump transfers heat from exterior to interior, heat can be moved between the earth and an interior space. Hot or cold water circulated through underground pipes will always return to the building at or near the earth's temperature.

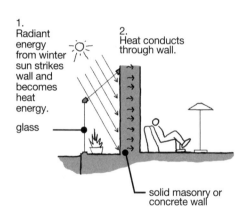

64

1.
Radiant energy from winter sun strikes wall and becomes heat energy.

glass

2.
Heat conducts through wall.

solid masonry or concrete wall

3.
In evening, heat radiating from wall modulates room temperature.

If wall is too thin, its thermal capacity will be reached too quickly, and heat will conduct to living space before it is needed.

A thermal storage wall

Available solar energy is 50,000 times our energy need.

At least 100 watts of energy strike each square foot of the earth's surface in a fully sunlit hour. Most areas of the U.S. receive the equivalent of 4 hours full sunlight per day, translating into about 1.5 trillion twh (terawatt hours) of energy per year—many times the 28,000 twh used in the U.S.

However, solar collectors currently capture only about 15 to 20% of the sun's energy (and are limited to a theoretical maximum of about 33%). And as the percentage of land that feasibly can be covered with solar collectors is small, it is difficult to meet all our energy demands through solar power. At present levels, the U.S. would need a continuous field of solar collectors covering the entire land area of Indiana. If the world used energy at the per capita rate of the U.S., a field the size of Venezuela would be required.

offset impervious
(hard) surfaces

reduce erosion by
minimizing rapid runoff

reduce load on
storm sewer system

support wildlife

reduce aesthetic
impact of development

Benefits of retention ponds

The environmental engineering paradigm shift

Responsibility is to stakeholders, not shareholders. Every creature and every part of the natural environment is a stakeholder in every project.

66

Ten Commandments for Environmental Engineers

1. Identify and promote the sustainable use of resources while maintaining a balance of social, economic, and environmental responsibility.
2. Provide safe, palatable drinking water.
3. Collect, treat, and discharge wastewater responsibly.
4. Collect, treat, and discharge human refuse responsibly to prevent disease, fire, and aesthetic insult.
5. Collect, treat, and discharge hazardous materials responsibly to prevent endangerment of human, plant, and animal life.
6. Control and treat air pollutants to reduce acid rain, ozone pollution, and global warming.
7. Design bioreactors to produce biofuels and electricity from organic waste.
8. Design physical, chemical, and biological processes to clean up contaminated sites.
9. Support and enforce legal regulation of the discharge of pollutants.
10. Study the fate and transport of chemical pollutants in air, water, and land.

67

With regards to Venkataramana Gadhamshetty

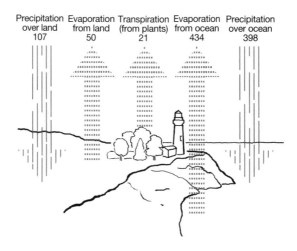

Precipitation over land 107 Evaporation from land 50 Transpiration (from plants) 21 Evaporation from ocean 434 Precipitation over ocean 398

Approximate annual global water movement in 1,000km³

Water is constant.

Water moves continuously below, on, and above the surface of the earth. Individual water molecules may come and go quickly or very slowly, but the overall amount of water remains fairly constant.

Reservoir	Average residence
Atmosphere	9 days
Soil	1 to 2 months
Seasonal Snow Cover	2 to 6 months
Rivers	2 to 6 months
Glaciers	20 to 100 years
Lakes	50 to 100 years
Groundwater, shallow	100 to 200 years
Groundwater, deep	10,000 years
Polar ice sheets	10,000 to 1,000,000 years

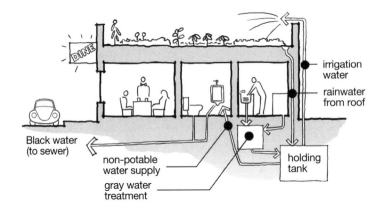

68

DINE

irrigation
water

rainwater
from roof

Black water
(to sewer)

non-potable
water supply

gray water
treatment

holding
tank

Gray water recycling

Water recycling

Black water has come into contact with fecal matter. It is not suitable for reuse without extensive treatment, typically via a municipal purification system.

Gray water is waste water from bathing, cooking, washing, and mild cleaning, and that has not come into contact with fecal matter. It is not safe to drink, but with some treatment is suitable for use in toilets and sometimes for the irrigation of plants.

White water is potable (consumable) water from a natural source such as a spring, or that has been treated by a municipal or similar purification system.

Rain water from outdoor surfaces such as roofs may contain some contaminants from birds and chemicals, but generally may be recycled for use in toilets, car washes, evaporative cooling systems, plant irrigation, and sometimes for consumption by livestock.

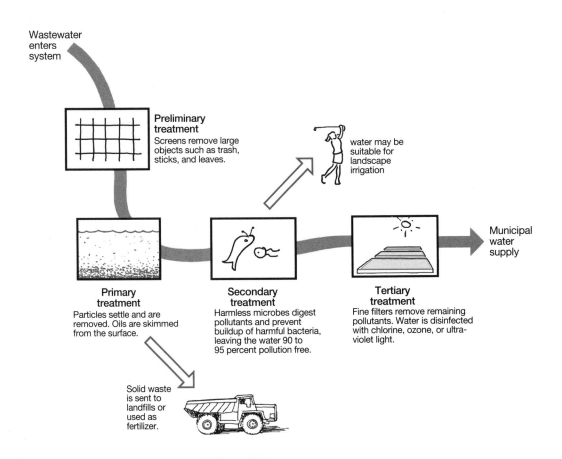

Wastewater enters system

Preliminary treatment
Screens remove large objects such as trash, sticks, and leaves.

water may be suitable for landscape irrigation

Municipal water supply

Primary treatment
Particles settle and are removed. Oils are skimmed from the surface.

Solid waste is sent to landfills or used as fertilizer.

Secondary treatment
Harmless microbes digest pollutants and prevent buildup of harmful bacteria, leaving the water 90 to 95 percent pollution free.

Tertiary treatment
Fine filters remove remaining pollutants. Water is disinfected with chlorine, ozone, or ultra-violet light.

Wastewater treatment imitates nature.

The complex systems used in wastewater treatment plants are accelerated simulations of nature's own purification processes:

- **Settlement basins** approximate a lake.
- **Filtration** approximates settlement to the groundwater table.
- **Aeration** approximates a stream.
- **UV treatment** approximates sunlight.

70

Our community needs
to build a bridge.

The design must
respect our
heritage as a
fishing harbor.

It should have minimal
environmental impact.

It must be in
harmony with
the setting.

TOWN HALL

The problem?

The solution

Don't presume the solution.

At the point a designer is invited into the design process, many assumptions have been made about the nature of a problem, its causes, and the desirable solutions. The wise designer begins by moving backward—investigating what caused the problem, what caused the causes, and what caused *those* causes. This reveals possibilities that might be very different from what the end user anticipated, but that meet the true need most effectively.

71

Proper deduction
specific conclusion
derived logically from more
general truths

Proper induction
multiple examples suggest,
but don't guarantee,
a larger truth

Improper induction
broad generalization or
declaration of similarity based
on limited data

Think systematically.

Don't congratulate yourself prematurely on an effective piece of analysis. Apply your thinking consistently and thoroughly to all other aspects of the problem at all possible scales, from concept to detail. Don't congratulate yourself then, either. There's more work to do.

72

Eames molded plywood chair

"We looked at the program and divided it into the essential elements, which turned out to be thirty-odd. And we proceeded methodically to make one hundred studies of each element. At the end of the hundred studies we tried to get the solution for that element that suited the thing best.... Then we proceeded to break down all logical combinations of these elements, trying to not erode the quality we had gained in the best of the hundred single elements; and then we took those elements and began to search for the logical combinations of combinations, and several of such stages.... And went right on down the procedure. And at the end of that time...we really wept, it looked so idiotically simple we thought we'd sort of blown the whole bit. And [we] won the competition."

—CHARLES EAMES, furniture designer,
from Ralph Caplan, *By Design*

73

3 single-span beams

2-span beam + single-span beam

One 3-span beam with cantilever

**Tension and compression behavior of beams
in three different structural arrangements**

Think *systemically*.

A system must be analyzed as a whole, but analysis of the whole is not the summation of the analyses of its parts. The behavior of a part is not constant, but depends upon its relationship to the system in which it resides. And the behavior of the system depends on the many relationships within it, and on the system's relationship to other systems.

Thinking systematically means employing a given thinking method consistently and thoroughly. Thinking *systemically* means thinking about systems and connections—the web of relationships within a system, the relationship of the system to other systems, and the larger system that contains all the systems.

74

Steel truss bridge
200' to 1800'

Cable stay bridge
500' to 3600'

Suspension bridge
1500' to 6500'

Approximate optimal main spans

A successful system won't necessarily work at a different scale.

An imaginary team of engineers sought to build a "super-horse" that would be twice as tall as a normal horse. When they created it, they discovered it to be a troubled, inefficient beast. Not only was it two times the height of a normal horse, it was twice as wide and twice as long, resulting in an overall mass eight times greater than normal. But the cross sectional area of its veins and arteries was only four times that of a normal horse, calling for its heart to work twice as hard. The surface area of its feet was four times that of a normal horse, but each foot had to support twice the weight per unit of surface area compared to a normal horse. Ultimately, the sickly animal had to be put down.

75

Deterministic system
outcomes can be predicted through
known laws and relationships

Stochastic/probabilistic system
outcomes depend on chance or
unknowable relationships

The behavior of simple systems and complex systems can be predicted. In-between systems: not so much.

The behavior of a single ball, set in motion by a known force on a billiard table, can be measured or predicted rather accurately. The individual behaviors of two balls are more difficult to measure or predict, but the task is somewhat manageable. But as the number of objects in the system increases—to five, ten, a hundred—tracking and/or predicting all the individual behaviors becomes extremely difficult and eventually impossible.

But at a later point, predictability re-enters the model, albeit in a different form. It is still difficult to reliably predict the individual behaviors of a million balls on an enormous billiard table, but we can predict many average behaviors, such as how many times a ball can be expected to hit another ball in a minute, how many balls will strike a particular rail in a second, or the average distance a ball moves before being hit by another ball.

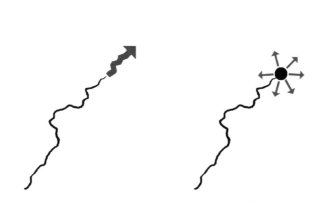

76

Stop a crack by rounding it off.

Crack propagation in a material increases with the sharpness of the tip of the crack. Drilling a hole at the tip makes a crack less sharp and distributes stresses over a larger area and in more directions, discouraging the crack from lengthening.

Rounded corners in building products, machine parts, furniture, and even the windows of ships and airplanes provide similar benefit. A rounded window corner spreads stress in multiple directions, while a sharply squared corner directs stress through one point in the system—a crucial consideration in the design of a "thin-shell" structure.

77

Thermostat
senses cold

Signal sent
to furnace

Temperature
decreases

Furnace
turns off

Furnace
turns on

I could have
drawn a
better furnace
than that.

Signal sent
to furnace

Temperature
rises

Thermostat senses
sufficient heat

Negative feedback

Seek negative feedback.

In a **negative feedback** loop, a system responds in the opposite direction of a stimulus, providing overall stability or equilibrium. For example, population growth of a species may result in overconsumption of the food supply; this leads to a decrease in population, which leads to an increase in available food, which promotes a population increase, and so on until a theoretical equilibrium is attained.

In a **positive feedback** loop, the system responds in the same direction as the stimulus, decreasing equilibrium further and further. For example, the introduction of a non-native species may invade the food supply of a native species; the native species retreats to outlying territory; this encourages the non-native species to expand its geographic range, leading to further retreat of the native species.

Nearly all engineered systems rely on negative feedback. In a few instances, such as when momentum is desired, a positive feedback loop may be sought.

78

The problem

|

the cause of the problem

|

the cause of the cause of the problem

|

the cause of the cause of the cause of the problem

|

the cause of the cause of the cause of the cause of the problem

|

the cause of the cause of the cause of the cause of the cause of the problem

|

the cause of the cause of the cause of the cause of the cause of the cause of the problem

|

the cause of the cause of the cause of the cause of the cause of the cause of the cause of the problem

Enlarge the problem space.

Almost every problem is larger than it initially appears. Explore and enlarge it at the outset—not to make more work, but because the scope of the problem almost certainly will creep—it will grow larger—on its own. It's easier to reduce the problem space later in the process than to enlarge it after starting down a path toward an inadequate solution.

79

3% grade
for trains

creek

25% grade—extreme maximum
for automobiles (equivalent to a
San Francisco street)

Tunkhannock Viaduct

The Tunkhannock Viaduct

The Delaware, Lackawanna and Western Railroad wished to replace a complex route between Scranton, Pennsylvania and Binghamton, New York with a straighter, flatter rail line. Key to its proposed Clark's Summit-Hallstead Cutoff was the crossing of the modest, 75-foot wide Tunkhannock Creek in Nicholson, Pennsylvania. As the creek was located in a valley too steep for trains to descend and climb, an otherwise modest crossing ended up requiring a 2,375-foot long, 240-foot high viaduct.

Construction began in 1912 and was completed in 1915. Thirteen piers were sunk to bedrock, the deepest twelve stories below the surface. The project used 1,140 tons of steel and 167,000 cubic yards of concrete—enough to cover a football field with a 7-story high solid mass. The viaduct was, and remained for at least 50 years, the world's largest concrete structure. It remains in daily use.

Chemists
investigate chemical interactions
and effects; create new solvents,
polymers, and pharmaceuticals

Chemical engineers
translate the discoveries of the
chemical laboratory into large-
scale industrial production

Almost everything is a chemical, and almost every chemical is dangerous.

Chemicals have a limited range in which they are useful to engineers. Below some thresholds of concentration, chemicals cannot be used productively in industry, and beyond a high limit of concentration they are toxic and unmanageable. Even water is a chemical, and it is dangerous in large amounts. Ingestion of too much can alter the body's chemical balance, deplete electrolytes, compromise organ functions, and cause death.

2H$_2$ + O$_2$ → 2H$_2$O

Balancing of hydrogen and oxygen atoms to make water

A chemical equation isn't exactly an equation.

A chemical equation does not represent equality in a pure mathematical sense, but serves to indicate the direction and result of a chemical reaction. When reactants are placed together, they interact to form a new compound, or product:

reactant + reactant → product

In the equation, "+" indicates "reacts with," and "→" indicates "produces."

A chemical equation *is* balanced in that atoms are not created or destroyed. The total number of atoms of each type before and after a reaction is constant, even though they are rearranged into new molecules.

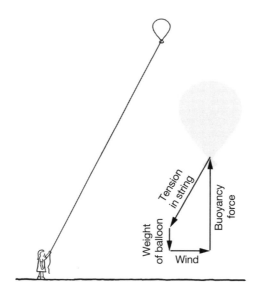

When an object is in equilibrium, the sum of all forces
acting on it is zero. The vectors form a closed polygon.

Equilibrium is a dynamic, not static, state.

When two chemicals come in contact and react, the reaction often appears to stop after a period of time as equilibrium is reached. Some portions of the chemicals will have combined into a new chemical product, while other portions appear unaffected. But even in equilibrium, the mixture often remains active, as portions of the product "uncombine" into the reactants and reactants continue combining into new product. However, the overall crossover rates balance and there is no net change in the system.

A structural equilibrium is similarly dynamic. A structural element, even though at rest, works quietly and unceasingly to resolve the various forces acting on it into an overall force of zero. Without the zeroing of forces, an object will accelerate, decelerate, or change direction.

83

Series circuit

Parallel circuit

Parallel circuit with
fuse wired in series

An electric current works only if it can return to its source.

A **series electrical circuit** employs a single loop that passes through each device it serves before returning to the source. Each device reduces the voltage available for the other devices; the greater the number of lamps, the more they will all dim. When one bulb burns out, the current path is interrupted and all the bulbs go out.

In a **parallel electrical circuit**, each device receives current from the power source without the current passing through other devices. The voltage is the same everywhere in the circuit, and the number of lamps does not affect their brightness. If one burns out, the others stay lit.

Because of the limitations of series circuits, parallel circuits are used to deliver electricity across cities and within buildings. However, fuses and circuit breakers work by being wired **in series within a parallel circuit**. When a fuse "blows" due to a power surge, it prevents damage to the other devices in the system by interrupting current flow in the same manner of a light bulb burning out in a series circuit.

**Moment =
2 ft. × 300 lbs. =
600 ft-lbs.**

**Moment =
−12 ft. × 50 lbs. =
−600 ft-lbs.**

A seesaw works by balancing moments.

A **moment** is a measurement of rotational tendency about a point, expressed by the equation:

Moment = force x perpendicular distance

The moment required to rotate a given object around a given point is constant, regardless of where the force is applied. A door, for example, may be opened by applying the necessary force a given distance from the hinge, or by applying twice the force at one-half the distance, or four times the force at one-fourth the distance. Force times distance is the same in all instances.

85

1. Suspend the object
from a point and mark
a vertical plumb line
from it.

2. Suspend it from a
second point and mark
a second plumb line.

Balance
point

3. The intersection
of the plumb lines is the
point on which it will
balance.

How to locate the center of mass of a flat, irregular object

Center of gravity

The center of gravity of an object is the average position of the particles that comprise it—the point on which it will balance. For objects of uniform density in an environment of uniform gravity, the **centroid** (geometric center), **center of gravity**, and **center of mass** are the same.

86

constant
diameter
shaft

receives
a nut

Bolt

unthreaded portion
helps compress
materials being joined

tapered shaft for
self-threading of
softer materials

Screw

It's called a *column*, not a *support column*.

And a beam is simply a beam, not a support beam; it's presumed it's there to provide support. Columns and beams are different things: columns are vertical and beams are horizontal. A column usually is made of steel, concrete, or masonry; if it's made of wood it's a post. There are no nucular engineers or masonary walls. A building foundation rests on a footing, not a footer. A sink in a kitchen is a sink, while a sink in a bathroom is a lavatory. A bolt and a screw are not the same thing, and by most standards a machine screw is misnamed; it's really more of a bolt. Steel isn't a pure metal, but an alloy; and stainless steel isn't stainless—it just stains less. A hot water heater is simply a water heater; water that's already hot doesn't have to be heated.

87

Articulate the *why*, not just the *what*.

When passing a design concept to other designers to develop, make known the reasons for the decisions made to that point, whether technical, ergonomic, personal, or other. By articulating your intent, you will encourage others to investigate alternatives that did not occur to you and help them preserve the most critical goals of the project. Perhaps most importantly, you will teach them that good design decisions have reasons.

Likewise, when a designer solicits your help in developing a project, ask for the reasons behind the decisions already made. This will reduce disappointment if you can't produce the specific result he or she expected.

88

John Fowler, Kaichi Watanabe, and Benjamin Baker, engineers of the
Firth of Forth Rail Bridge, demonstrate its structural system in 1887.

All engineers calculate. Good engineers communicate.

The scientific concepts, analytical processes, and mathematical calculations used to solve engineering problems have been developed over hundreds of years. Early in this period, engineers came to speak a common language based in mathematics, chemistry, and physics, such that any engineering solution developed anywhere in the world could be read and understood by effectively every engineer.

As more engineering specialties have emerged, engineering "dialects" have arisen. Engineers now must be aware of highly specialized terminology and concepts within their discipline, and must be ever more able to translate their knowledge and ideas into everyday language that can be understood by clients, end users, and other engineers.

89

Completed form

back span · cantilever **+** simple span
supported at ends
by cantilevers **+** cantilever · back span

How it works structurally

Common double cantilever truss bridge

How to read, but not necessarily name, a cantilever bridge

Most large bridges are built by the **cantilever method**. Concrete piers are constructed in the river or chasm, and from each, structural members are increasingly extended (cantilevered) in opposite directions. This allows overall loads to remain balanced, and construction to proceed from the safety of built areas toward unrealized portions of the span. Eventually the cantilevers meet, either directly or via an additional structural component.

In a steel truss bridge, if the cantilever construction method is used, it usually functions as a cantilever structure upon completion and is permanently classified as a cantilever bridge. Sometimes the cantilevers "read through" as distinct elements, as in the Firth of Forth Rail Bridge in Scotland, while in most bridges they are less distinct.

Cable stay, suspension, and even some concrete arch bridges also are constructed by variations of the cantilever method. However, when completed they do not function structurally as cantilever systems and are not classified as cantilever bridges. Conversely, some steel truss bridges built by the cantilever method do not behave as pure cantilever structures upon completion, but *are* permanently classified as cantilever bridges.

Painter

Musician

Technical   Author
writer

Fashion   English
designer   teacher

Computer
programmer   Clergy

Engineer   Chemistry
                        teacher
Banker   Therapist

 Machinist   Merchant

Random hypothesis #2

There are three kinds of people in the world: language people, people people, and object people. Language people find meaningful connections with the world through written, spoken, and symbolic communication. People people seek empathic connections to other people and to human causes. Object people experience the world primarily through relationships with physical things. But object people don't merely "like" objects; they understand the world from the viewpoint of the objects with which they are concerned.

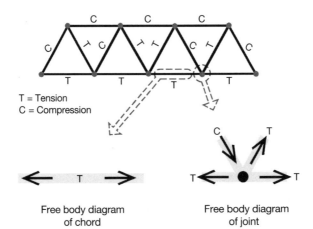

T = Tension
C = Compression

Free body diagram
of chord

Free body diagram
of joint

Right now, you're the chord. In a minute, you'll be the joint.

When struggling to analyze a complex problem, shift your point of view from that of outside observer to that of the thing you are analyzing. If you were that thing, what forces would you feel? What internal stresses would you experience? How would you have to react in order to remain stable and not twist, turn, deform, be pushed over, or caused to accelerate?

Structural analysis of trusses requires that one continually shift point of view among its chords and joints in this way. If you don't, you can get the force arrows pointing the wrong way. Where a chord is in tension, for example, the force arrows point outward from it as one might expect. But at the joint to which the same chord connects, the force arrows point the opposite way—back toward the chord. This is because tension pulls on *all* the components involved, but each component experiences the tension from its particular point of view.

$$\text{Satisfaction} = \frac{\text{outcome}}{\text{input}}$$

Activity satisfaction

$$\frac{\text{outcome}}{\text{input}} \longleftrightarrow \frac{\text{outcome}}{\text{input}}$$

Activity comparison

$$\frac{\text{outcome}}{\text{input}} \longleftrightarrow \frac{\text{peer's outcome}}{\text{peer's input}}$$

Situational comparison

$$\frac{\text{outcome}}{\text{input}} \longleftrightarrow \frac{\text{partner's outcome}}{\text{partner's input}}$$

Relationship comparison

The engineering of satisfaction

The satisfaction one derives from an activity can be roughly quantified as the ratio of reward to input. Reward does not have to exceed input for a person to feel an endeavor is worthwhile. Rather, a favorable comparison of one's situation to other possible situations or to a peer's situation is the more crucial determinant. When people feel fairly or advantageously rewarded—when their ratio is at least as high as a peer's ratio—they are more likely to be motivated. When people feel unfairly rewarded, they are inclined toward disinterest, demotivation, and resentment. When over-rewarded, feelings of guilt may result.

In relationships of an interpersonal nature, satisfaction likely will be mutual if the partners experience equal ratios, regardless of whether the amounts each gives and receives are equal to the amounts the partner gives and receives.

suspenders
(vertical)

cable stays
(diagonal)

main cable

stiffening truss

anchorage

Engineering events are human events.

A beloved symbol of American ingenuity and optimism, the Brooklyn Bridge had an inauspicious beginning. Engineer John Augustus Roebling's foot was crushed by a ferry while surveying the location. After having several toes removed, and while dying from tetanus, he appointed his 32-year-old son Washington to take over the project. Less than three years later, the younger Roebling exited a pressurized excavation chamber too quickly and fell ill. Nearly paralyzed, he was confined to his apartment for the next eleven years. His wife Emily supervised the work to completion.

In May 1883, after the death of more than two dozen workers, the longest suspension bridge in the world and the first to use steel wire opened. Washington Roebling was unable to attend. Emily crossed first, followed by 1,800 vehicles and 150,000 pedestrians.

Some New Yorkers were suspicious of the bridge's strength, however. It was fully 1½ times the length of any previous suspension bridge, and it had been discovered during construction that the cabling contractor provided inferior wires. At the time, Roebling had 250 additional cables installed diagonally from the towers to the deck, turning the bridge into a hybrid suspension/cable-stayed structure and giving it its uniquely elegant, cobwebby look. But concerns lingered. Several days after the opening, a woman on the bridge screamed, sending hundreds into a panicked stampede to exit the bridge. Twelve were killed.

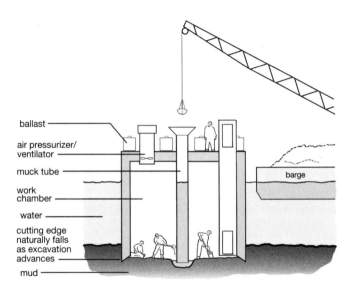

ballast

air pressurizer/
ventilator

muck tube

work
chamber

water

cutting edge
naturally falls
as excavation
advances

mud

barge

Bridge caisson cross section

There's design besides the design.

A well-designed product *isn't* well designed if the process needed to manufacture it is unrealistic or uneconomical. A brilliantly conceived, alternative-fuel vehicle will not succeed without the design and implementation of a refueling infrastructure over a large geographic area. A cleverly resolved construction detail isn't clever if it doesn't leave room for a construction worker to manipulate the tools needed to construct it. A bridge pier that's well engineered won't be built without also engineering a process to excavate earth and pour concrete in the middle of a river.

95

PRIORITY

"The most important thing is to keep the most important thing the most important thing."

—DONALD P. CODUTO, *Foundation Design*

97

Bridge tower: provide nesting sites for hawks, eagles

Highway barrier: gabion wall construction uses scrap rock, encourages climbing vegetation, discourages graffiti

Road overpass: provide shelves underneath for bats

Wastewater treatment: methane off-gas used to power operations

While getting the one thing right, do more than one thing.

Engineering is a field of specialties, and engineers are called on to solve specific problems. Don't get distracted by all the other possibilities such that you forget to do the one thing you must do. But don't become so focused on the one thing that you don't do as much as you can.

98

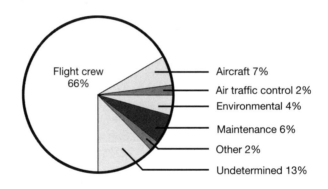

Worldwide causes of airplane accidents

The fix for an apparent engineering problem might not be an engineering fix.

On March 27, 1977, a Pan Am 747 was taxiing on the runway of Tenerife North Airport in the Canary Islands as a KLM Airlines 747 attempted to take off. The resulting collision killed 583 people, the most ever in an aviation accident. Numerous physical contributors were identified, including:

- unusually high traffic due to the temporary closing of a larger airport nearby
- many airplanes parked on the taxiway, complicating taxiing patterns
- dense fog that greatly limited visibility
- no ground radar; controllers had to rely on radio to identify plane positions
- simultaneous radio transmissions that cancelled each other, causing messages to be unheard or misheard, and leading to unauthorized takeoff by the KLM captain despite concern from the co-pilot

Among the industry-wide changes after the disaster:

- The word "takeoff" was forbidden except when the control tower authorizes an aircraft to take off. At all other times, "departure" or another term is to be used.
- Flight crew across the industry were retrained, with lower ranking crew encouraged to challenge captains with any concerns, and captains required to consider crew concerns in making all decisions.

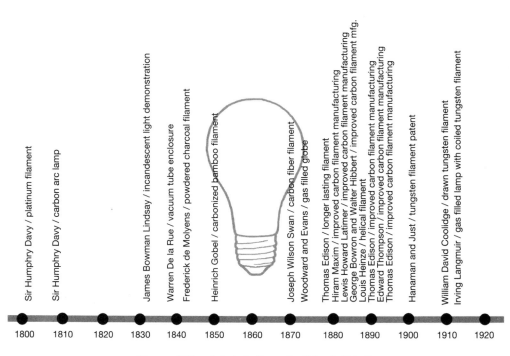

99

Thomas Edison's help in "inventing" the light bulb

Timeline labels:

- Sir Humphry Davy / platinum filament
- Sir Humphry Davy / carbon arc lamp
- James Bowman Lindsay / incandescent light demonstration
- Warren De la Rue / vacuum tube enclosure
- Frederick de Molyens / powdered charcoal filament
- Heinrich Gobel / carbonized bamboo filament
- Joseph Wilson Swan / carbon fiber filament
- Woodward and Evans / gas filled globe
- Thomas Edison / longer lasting filament
- Hiram Maxim / improved carbon filament manufacturing
- Lewis Howard Latimer / improved carbon filament manufacturing
- George Bowron and Walter Hibbert / improved carbon filament mfg.
- Louis Heinze / helical filament
- Thomas Edison / improved carbon filament manufacturing
- Edward Thompson / improved carbon filament manufacturing
- Thomas Edison / improved carbon filament manufacturing
- Hanaman and Just / tungsten filament patent
- William David Coolidge / drawn tungsten filament
- Irving Langmuir / gas filled lamp with coiled tungsten filament

1800 1810 1820 1830 1840 1850 1860 1870 1880 1890 1900 1910 1920

Engineering usually isn't inventing the wheel; it's improving the wheel.

Great inventions usually are incremental steps forward from previous achievements. Gideon Sundback invented the modern zipper in 1917. However, Whitcomb Judson developed the "Clasp Locker" eighteen years earlier and Elias Howe, inventor of the sewing machine, patented an "Automatic, Continuous Clothing Closure" more than 40 years before that. Sundback replaced the hook-and-eye fasteners of these earlier devices with "scoop-dimpled" teeth, greatly increased the number of fasteners per inch, and provided the familiar slide mechanism that opens and closes the system. Six years after Sundback patented his "Separable Fastener," B. F. Goodrich coined the onomatopoeic "Zipper" when his company introduced galoshes with Sundback's device.

100

Philosopher

contemplates
paradigms,
meaning, value
of human
endeavors

Scientist

identifies
principles of
nature through
hypothesis and
experiment

Engineer

designs useful
things based
on proven
scientific
principles

Technician

inspects,
troubleshoots,
and implements
using known
methods

User

desires seamless
application;
usually has little
technical
knowledge

The Great Continuum

Engineering is undertaken within a continuum that connects profound human questions to ordinary activities. Engineers who work without awareness of the continuum will be inclined toward performing rote procedures. Those working in awareness of it will be better positioned to adapt to changing times, unexpected challenges, and unfamiliar circumstances. Those working across the continuum may be most apt to contribute something new.

Notes

Lesson 2: Adapted from Mark Holtzapple, W. Reece, *Foundations of Engineering* (McGraw-Hill Science/Engineering/Math, 2nd ed., 2002), p. 9

Lesson 5: With regards to Ralph Caplan, *By Design*

Lesson 52: Adapted from Frederick Gould, *Managing the Construction Process* (Prentice Hall, 4th ed., 2012), p. 64

Lesson 53: Gary T. Schwartz, "The Myth of the Ford Pinto Case," *Rutgers Law Review*, Vol. 43, p. 1029

Lesson 67: Adapted from John Elkington, *Cannibals with Forks*

Lesson 75: Derived from "The Possibility of Life in Other Worlds" by Sir Robert Ball, *Scientific American Supplement*, No. 992, January 5, 1895, pp. 15859–61

Lesson 76: With regards to the late Jane Jacobs, *The Death and Life of Great American Cities*

Lesson 99: "Global Fatal Accident Review, 1997-2006," UK Civil Aviation Authority

Index

John Kuprenas is a registered engineer and LEED professional in California. He lectures in civil engineering at the University of Southern California and at California State University, Long Beach, and is vice president at Vanir Construction Management in Los Angeles. His writings have been published in numerous refereed journals and in *The Story of Managing Projects* (Greenwood Press/Quorum Books).

Matthew Frederick, an architect, is the author of *101 Things I Learned® in Architecture School* and the creator, editor, and illustrator of the 101 Things I Learned® series. He lives in Hudson, New York.